Gleison Siqueira de Souza
Ademar Paulo Junior

Theoretical study of semiconductor materials

Gleison Siqueira de Souza
Ademar Paulo Junior

Theoretical study of semiconductor materials

Technological applications

ScienciaScripts

Imprint

Any brand names and product names mentioned in this book are subject to trademark, brand or patent protection and are trademarks or registered trademarks of their respective holders. The use of brand names, product names, common names, trade names, product descriptions etc. even without a particular marking in this work is in no way to be construed to mean that such names may be regarded as unrestricted in respect of trademark and brand protection legislation and could thus be used by anyone.

Cover image: www.ingimage.com

This book is a translation from the original published under ISBN 978-613-9-66693-5.

Publisher:
Sciencia Scripts
is a trademark of
Dodo Books Indian Ocean Ltd. and OmniScriptum S.R.L publishing group

120 High Road, East Finchley, London, N2 9ED, United Kingdom
Str. Armeneasca 28/1, office 1, Chisinau MD-2012, Republic of Moldova, Europe
Printed at: see last page
ISBN: 978-620-8-05822-7

I dedicate this work to my family and my dear and loving wife Millena, for always being by my side.

ACKNOWLEDGEMENTS

"But those who hope in the Lord shall renew their strength, and mount up with wings like eagles; they shall run and not be weary, they shall **walk and not faint."** (Isaiah 40:31)

I thank God for granting me this blessing, giving me health, perseverance and the certainty of victory, not letting me lose heart even in difficult times.

My family, especially my mum, a strong woman, a warrior and an example to me, who always supported me and believed in my success, being a firm base of support.

To Millena, friend, girlfriend, fiancée and now my beloved wife, for her encouragement, for being by my side, being my strong arm, contributing to this work.

To my father-in-law, mother-in-law and sister-in-law for welcoming me into the family with open hearts and for being part of my life.

To the pastor and brothers in Christ Jesus of the First Baptist Church in Taquaruçu for their prayerful support.

To my supervisor Ademar, for his courage and strength of will in making a significant contribution to this work.

To the teachers who were part of my academic training, especially Weimar, Denise and Márcio.

To friends, colleagues and everyone who has passed through my life and has contributed in some way to my academic and personal growth.

"[..]] The good teacher is the one who manages, while speaking, to bring the student into the intimacy of the movement of his thought. His lesson is thus a challenge and not a lullaby. Their students get tired, they don't sleep. They get tired because they follow the comings and goings of their thinking, they are surprised by their pauses, their doubts, their uncertainties. " Paulo Freire

SUMMARY

SOUZA, Gleison Siqueira. Theoretical study of semiconductor materials and technological applications. 2014. 65f. Course Conclusion Programme. Federal Institute of Education, Science and Technology, Palmas Campus, Palmas -TO, 2014.

This work is a theoretical study that presents a little of the history of atomism, from the first idea of the atom in ancient Greece, through some important authors such as Dalton, Thomson and Rutherford, to Bohr's model, which is not the current model, but which played an important role in understanding the microscopic world by determining the atom's energy levels. From Bohr's model, we present semiconductor materials and their particularities, which, thanks to the knowledge of energy levels, can be worked on to better fulfil certain purposes. These semiconductor materials are the basis of technological advances with many applications in society. Knowledge of the electronics of these materials, which has come about through the evolution of physics, has led to improvements in people's daily lives. These applications of physics should be worked on in the classroom with students who often don't see a real reason to study physics. Physics has evolved and improved people's well-being. Teachers should be aware of this and know how to work with students on physics applications so that they can enjoy the subject.

Keywords: semiconductors; physics; applications; atoms.

Physics knowledge has evolved and brought improvements to human life, and this work aims to show the relationship between physics and everyday life, since in the classroom students who don't understand this reality end up disliking the subject, making learning difficult. To this end, we present the concept of the atom from Greece to Bohr and semiconductor materials with their applications, which can be used as examples to arouse students' curiosity or answer their questions.

The first idea of the atom emerged in Greek philosophy with the philosophers Leucippus of Miletus and Democritus of Abdera. For them, the atom would be the smallest part of matter, which does not divide, arising from man's need to explain nature. With the various thoughts of the time, and the lack of experimentation to prove the idea in order to consolidate a new theory, this thought was overtaken by other relevant ideas of the time.

It wasn't until the 19th century that the English physicist John Dalton came up with the first atomic model. From then on, the race to understand the structures of the atom became more intense. At the end of the 19th century, with the so-called sultana pudding model, Joseph John Thomson presented his idea of the atom, a solid sphere with the electrons in its shell, which until then had been the first atomic model divisible into parts. Thomson's former student Ernest Rutherford revealed the existence of the atomic nucleus surrounded by electrons and made up mostly of empty space. In 1911 Rutherford carried out the famous gold leaf experiment, resulting in a model in which the electrons revolved around the nucleus. In Rutherford's model, the electrons revolved around the nucleus, like the solar system.

However, this new model left an intriguing question: how did these electrons revolve around the nucleus and not collide with it? Rutherford's model was unable to answer this question. Niels Henrik David Bohr, who worked in Rutherford's laboratory, dedicated himself to answering why the electrons orbited the atomic nucleus without collapsing. Bohr believed that, with Max Planck's quantum theory, he would be able to answer this question and create a new model. In 1913, Bohr presented his atomic model for the hydrogen atom. Using this atomic model, with the electrons revolving around the nucleus, Bohr determined the atom's energy levels.

With Bohr's atomic model and the understanding of energy levels, it was possible to learn about the electrical behaviour of materials, including semiconductors, with electrical conduction intermediate between insulators and conductors. This type of material can be engineered to better serve certain purposes. The study of semiconductor electronics has led to major changes in technology. The applications of semiconductor devices are present in new technologies, making them easier to handle, saving energy and miniaturising their dimensions.

In the first chapter of this work, we will briefly discuss the history of atomism from ancient Greece to Bohr's atomic model, with a greater emphasis on the latter, as it made a major contribution to the study of semiconductors. The second chapter deals with semiconductors, how they conduct electricity and their physical structure. The third chapter will focus on semiconductor devices and their technological applications.

CHAPTER 1

ATOMISTIC HISTORY, ATOMIC MODELS

Chemists define matter as everything that has mass and occupies a place in space. When portions of this matter, its smaller parts and its behaviour are known, it is possible to work on them for the benefit of man. With a deeper understanding of the atom, it was possible to understand its behaviour and, with this, to manipulate certain substances, driving technological progress. Ferreira, Bampi and Alvares (2004) agree that technological development took place through semiconductor electronics, thus making it possible to advance information technology, communications, revolutionising people's daily lives and business.

However, the beginning of the search for knowledge of the nature of matter can be traced back to ancient Greece, around the 5th century BC, where the atom was conceptualised as the smallest part of matter, that which does not divide, from the Greek (a = not; tomos = does not divide). In the following sections, we will briefly review atomistic theory from its origins to Niels Henrik David Bohr's atomic model in 1913, where its definition led to major scientific advances, one of the most important being the discovery and technological application of semiconductors, Ferreira, Bampi and Alvares (2004).

1.1 Greek atom

The idea of the atom arose in Greek philosophy with Leucippus and Democritus of the atomist school. Contrasting the doctrines of the philosophers Parmenides of Elaea, Zenon of Elaea and Melissus of Samos of the school of Elaea, they affirm that bodies are composed of atoms, indivisible, independent and immutable beings, as we can see Reale and Antiseri (1990) affirm that their doctrines and philosophies are due to man's need to explain the environment, natural phenomena and the origin of the world.

According to Reale and Antiseri (1990) Leucippus is the founder of the atomist school. Born in Miletus, he moved to the city of Elea in Italy where he learnt about the Eleatic doctrine. He later moved to Abdera, where he founded the atomist school, whose disciple was Democritus, born around 460 BC, who is said to have raised the idea of the atom.

> The atomist doctrine holds that reality consists of atoms and emptiness, with atoms attracting and repelling each other, thus generating natural phenomena and movement. The attraction and repulsion of atoms is due to their geometric shapes, with atoms of similar shapes attracting each other and those of different shapes repelling each other. (MARCONDES, 2001, p. 34-35)

With the advent of the sophist movement, men with unusual eloquence, wise orators who had the power of persuasion, who fascinated those who listened to them, with Protagoras, a cultured man as the first sophist, and the philosophers Socrates, Plato and Aristotle, atomistic thinking lost ground in Greek philosophy. Atomistic philosophy was revived by Epicurus of Samos (~342-271 BC) and Lucretius (Rome 100/94-55 BC). According to Filgueiras (2004), the opposition of Aristotle, considered the main philosopher in the Middle Ages and Renaissance, made atomic theory unacceptable and heretical in the eyes of the Western Church.

The concept of the atom remained in the philosophical realm until around the beginning of the 19th century, when John Dalton presented his atomic model, the first scientific concept of the atom.

1.2 John Dalton's atomic model

John Dalton (1766-1844), Figure 1, was born in Eaglesfield, England, did not go on to university, but always showed an interest in maths. From 1784 to 1794 he wrote in periodicals, studied zoology and botany, made meteorological observations and taught courses in natural philosophy and the chemistry of gases, which would lead him to the atomic model. In 1793 he was invited to teach maths and natural philosophy at New College in Manchester (FILGUEIRAS, 2004). According to Viana and Porto (2007), Dalton, persuaded by Newton's corpuscular idea in the late 18th and early 19th centuries, adhered to this way of thinking about matter. In addition, his search for explanations for issues related to the atmosphere and gases led him to speculate about possible differences in mass between the atoms that make up the different elementary substances.

Figure 1 - John Dalton.
Source: Just chemistry.

According to Filgueiras (2004), Dalton, who was passionate about meteorology, began to study the gases that make up the atmosphere. At the time, researchers wondered how these gases - oxygen, nitrogen and carbon dioxide - interacted with each other. From then on, he set out to find answers through experiments based on Newtonian principles and ended up discovering an important law, the law of partial pressures of gases.

With this theory, Dalton did not stop, and sought to explain the interdependence of one gas in relation to another, considering whether it was due to the difference in particle size. This is how he began his study with dissolution in water, bringing his milestone to modern chemistry and the history of the atom, the law of multiple proportions, which gave rise to all the calculations of the mass of atoms (FILGUEIRAS, 2004).

For the time, the theory of the weights of atoms took a while to be accepted, as everything Dalton proposed was very new. However, as the years went by, the idea of the atomic model was accepted and for a long time it answered the scientific doubts and questions of the time, lasting until the end of the 19th century when the first signs that the theory was outdated began to appear (MELZER and AIRES, 2011).

"In fact, Dalton resurrected the concept of the Greek atom and was able to support this concept with experimental evidence that he and others obtained." (RUSSELL, 1994, p. 207). With his work, and represented in Figure 2, Dalton concluded in 1803 that:

All matter is made up of particles called atoms;

- Atoms are indivisible, they can neither be created nor destroyed according to the principle of conservation of matter - Lavoisier;

- Atoms of the same element are identical and have the same weight;

- This model showed the atom as a massive sphere, and became known as the billiard ball model.

Figure 2 - Dalton's atomic model.

Source: Explicatorium.

Over time, this model stopped answering some questions. In the second half of the 19th century, experiments showed mistakes, such as the atom not being a massive sphere, not being indestructible, being divisible, penetrable and sometimes different in the same chemical element.

1.3 Joseph John Thomson's atomic model

Joseph John Thomson, Figure 3, was a British physicist who in 1897 discovered the electron, and later the ratio between its charge and mass. According to Eisberg and Resnick (1979) in the first decade of the 20th century there was already a lot of experimental evidence for the existence of negatively charged particles, later called electrons, such as X-ray scattering, the photoelectric effect and others. "These experiments gave an estimate for Z, the number of electrons in an atom, as being approximately equal to A/2, where A is the chemical atomic weight of the atom considered." (EISBERG and RESNICK, 1979, p. 123).

Figure 3 - Joseph John Thomson.
Source: Fermi national accelerator laboratory.

Young and Freedman (2003) report that these experiments, as well as others that are not cited in this work, showed that the positive charge corresponds to most of the total mass of the atom, and not the negative charges, the electrons. Therefore, **"the fact that the mass of the electron is** very small compared to that of any atom, even the lightest, implies that most of the mass of the atom must **be associated with the positive charge"** (EISBERG and RESNICK 1979, p. 123).

According to the authors mentioned above, considerations about the distribution of charges within the atom and experiments such as electrical discharges in Crookes tubes led Thomson to discover the existence of the electron. Crookes tubes are basically a device made up of a glass ampoule connected to a vacuum pump containing two electrodes at the ends. This work led J. J. Thomson to propose his model of the atom at the end of the 19th century. In this model, the negatively charged electrons would be inside a continuous, spherical distribution of positive charges. The radius of this sphere was already known and was in the order of 10^{-10} m. This atomic model became known as the sultana pudding because it looked like a pudding filled with sultanas, as shown in Figure 4.

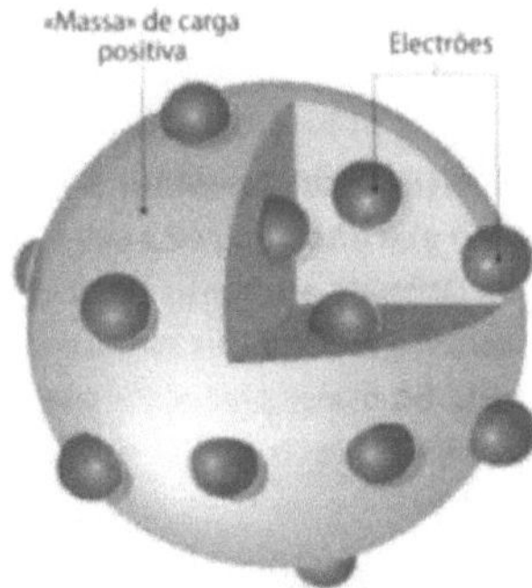

Figure 4 - Joseph John Thomson's atomic model.
Source: Atoms chemistry group.

In this model, when the atom was in its ground state, the lowest energy state, the electrons would be fixed in an equilibrium position, and in excited atoms, the electrons would vibrate around this equilibrium position. According to electromagnetic theory, this atom would emit radiation because it is a charged body and is in accelerated motion. With this theory it was possible to understand the emission of electromagnetic radiation from excited atoms through the Thomson model, but it didn't agree with the line spectra observed experimentally. (EISBERG and RESNICK, 1979)

This information was already known to physicists at the time, the main question being the behaviour of these charges within the atom. At the beginning of the 20th century, Ernest Rutherford and his collaborators developed a new model of the atom, the nuclear atom, based on the surprising results obtained with the alpha particle scattering experiment.

1.   4Ernest Rutherford's atomic model

Born in New Zealand, Ernest Rutherford (1871-1937), Figure 5, developed his professional activities in England and Canada. Together with Frederick Soddy, he won the Nobel Prize for Chemistry in 1908 for showing radioactivity in the disintegration of atoms, Young and Freedman (2003).

Figure 5 - Ernest Rutherford.
Source: University of Canterbury.

According to Russell (1994), the discovery of radioactive elements enabled Rutherford and his collaborators, Johannes (Hasn) Wilhelm Geiger and Ernest Marsden, to carry out an experiment using alpha particles, positively charged particles with a mass much greater than that of an electron, around 7300 times. Today we know that alpha particles are actually the nuclei of helium atoms made up of two protons and two neutrons.

Rutherford's experiment consisted of bombarding thin sheets of various materials, including gold, and observing the behaviour of the particles, which either passed through or were deflected.

The particles were detected on fluorescent zinc sulphide screens, where a flash formed at the point of contact between the particle and the screen, as shown in Figure 6.

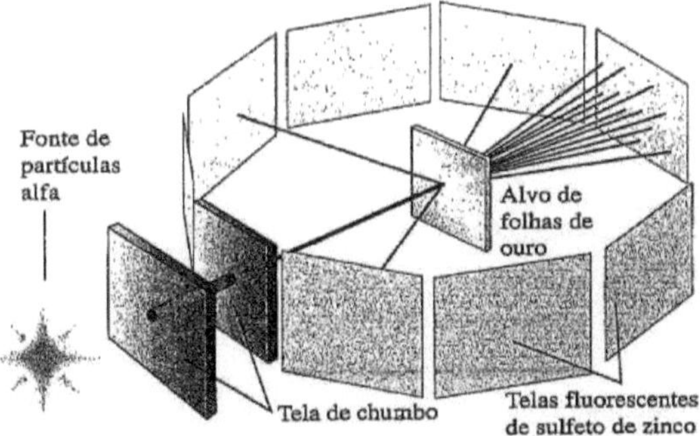

Figure 6 - Rutherford's experiment, alpha particle scattering.
Source: Young and Freedman.

A source of radioactive substance emits alpha particles, and two thick lead screens with a central hole collimate the particles into a narrow (or thin) beam. The beam collides with the target, the gold leaves, and it is observed that many particles pass straight through with little or no deviation and others are deflected, even returning towards the radioactive source. Rutherford's idea was to measure the deviation of the alpha particles as they passed through the atoms of the material.

When Rutherford's experiment was carried out, the model of the atom proposed by J. J. Thomson had all the credibility and the electrons would be distributed in a positively charged sphere, as stated by Halliday, Resnick and Walker (2011). However, this model was not satisfactory for explaining what was observed in Rutherford's experiment. It explained the passage of the alpha particles with small deviations, but did not explain the large deviations of the particles, as shown in Figure 7.

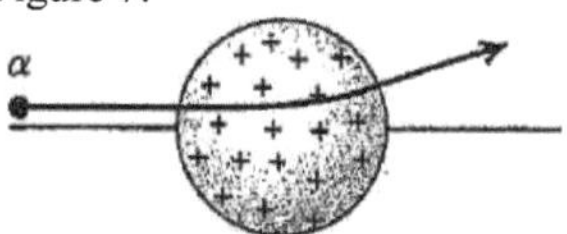

Figure 7 - Scattering of alpha particles according to Thomson's model.

Source: Young and Freedman.

According to Thomson's model, the alpha particles would suffer small deviations as they passed through the atom, due to their large size in relation to the electrons.

"Thinking in terms of Thomson's model, at first he [Rutherford] was not surprised by the fact that many of the alpha particles travelled in a straight line across **the sheet with little or no deflection"** (RUSSELL, 1994, p. 215). However, the question remained, what about the large deviations? as shown in Figure 8.

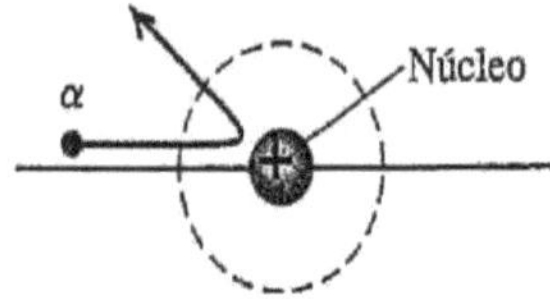

Figure 8 - Facts observed in the scattering of alpha particles.
Source: Young and Freedman.

As we see in Halliday, Resnick and Walker (2011), Rutherford imagined that the particles that suffered large deviations had to be subjected to a very large force, and for this to happen, the positive charge, instead of being scattered in the atom, had to be concentrated in a tiny space. The maximum proximity of the alpha particles would result in a very strong force of repulsion, resulting in large deviations, even with an angle of 180°.

With this experiment Rutherford presented his nuclear atomic model in 1911, in which the atom has a small nucleus, with a nuclear radius of around IO^{-14} m containing approximately 99.95

% of the total mass and occupying a volume around 10^{-12} times the volume of the atom, where it concentrates all the positive charge, with a size 10^4 times smaller than that proposed by Thomson. The electrons described orbits around the nucleus, figure 9, analogous to the planetary system, distributed in a very large space called the electrosphere, as shown by the authors Eisberg and Resnick, (1979), Halliday, Resnick and Walker, (2011) and Young and Freedman, (2003).

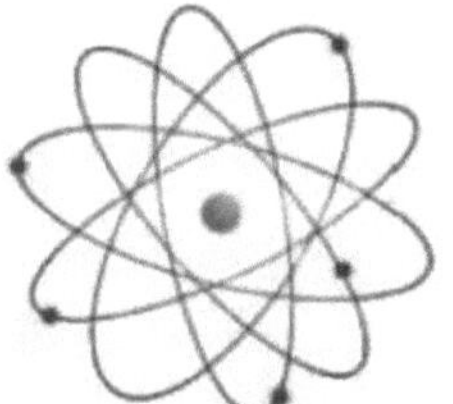

Figure 9- Rutherford's atomic model.
Source: Military College of Fortaleza.

In 1914, Rutherford also demonstrated the existence of a particle with a much greater mass than the electron and a charge of the same magnitude, but with the opposite sign, later called protons. Protons alone were not enough to explain the constitution of the nucleus, and in 1932 English physicist James Chadwick detected the existence of neutrons, particles of the same mass as protons but with zero electric charge, in the nucleus of the atom. In this way, the atom is represented as follows: a central nucleus made up of positively charged protons containing most of the mass and occupying a small space, neutrons with no electrical charge and negatively charged electrons with orbits occupying a large volume around the nucleus. The net charge of an atom in its fundamental state is zero, due to the fact that the number of protons is equal to the number of electrons, and their charges are of opposite signs. Russell (1994).

1.5 Niels Henrik David Bohr's atomic model

Rutherford's atomic model, as discussed above, showed the electrons gravitating around the nucleus, which occupies a very small space with a high concentration of mass, but it didn't give an answer as to the stability of this atom.

Niels H.D. Bohr (1885-1962), Figure 10, a Danish physicist, proposed a model for the hydrogen atom, which is the simplest atom containing a proton and an electron. His model, supported by the work of Max Planck, Albert Einstein and Rutherford, predicted the position of the lines in the hydrogen spectrum (TIPLER and LLEWELLYN, 2010). Bohr's proposal arose from the need to explain the inconsistencies between Rutherford's atomic model and classical theory.

Figure 10 - Niels H. D. Bohr.
Source: Just chemistry.

According to these authors, Bohr thought of the hydrogen atom's electron rotating in a circular orbit around the nucleus. In this model, the Coulombian interaction force guarantees mechanical stability, because the resultant is a centripetal force, which keeps the electron rotating with speed v around the nucleus.

$$F = \frac{kZe^2}{r^2} = \frac{mv^2}{r} \tag{01}$$

where e is the elementary charge of the electron. On the other hand, the model would be electrically unstable since, according to Maxwell's laws of electrodynamics, a charge in accelerated motion radiates energy by means of electromagnetic waves and the electron, rotating around the nucleus, would be accelerated in its direction radiating energy at a frequency equal to the frequency of revolution. We know that the frequency of revolution is given by

$$f = \frac{v}{2\pi r} \tag{02}$$

where r is the radius of the orbit and v is the speed of the electron. By isolating v in equation (01) and substituting it into (02), we arrive at the equation for the frequency of revolution as a function of radius

$$f = \left(\frac{kZe^2}{4\pi^2 m}\right)^{\frac{1}{2}} \cdot \frac{1}{r^{\frac{3}{2}}} \sim \frac{1}{r^{\frac{3}{2}}} \tag{03}$$

Therefore, this electron would continually lose energy. As we know, the electron's total energy is given by the sum of its kinetic and potential energy, as shown in the equation

$$E = \frac{1}{2}mv^2 + \left(-\frac{kZe^2}{r}\right) \tag{04}$$

However, due to the loss of energy, the radius of the trajectory would decrease and the electron would describe a spiral orbit until it collided with the nucleus. It is possible to calculate the time needed for the electron to reach the nucleus of the atom, but we won't do this calculation as it is not the aim of this work.

However, the electrons do not collapse into the nucleus, as theoretically predicted, since the hydrogen atom is stable, as is matter. The experiments showed that atoms don't radiate energy simply because the electrons are orbiting the nucleus in accelerated motion, but this requires external action, because the atoms don't radiate a continuous spectrum, but only well-defined frequencies.

Classical physics can no longer provide a theoretical explanation for the phenomena observed experimentally. For Russell (1994) there are two possibilities to consider: either it is stationary or it is moving. The electron being stationary poses the following dilemma according to the principles of classical mechanics: the force of attraction between the positively charged nucleus and the negatively charged electron would bring them closer together until they collapsed, making it impossible for any kind of matter to exist. On the other hand, with the electron in motion, as mentioned above, the end would be the same as in the previous situation, and the electrons would emit a continuous spectrum of radiation, different from the knowledge that scientists at that time already had about discrete spectra emitted by atoms.

These conclusions reached at the beginning of the 20th century, as Russell (199) points out, became a dilemma for the scientists of the time. The explanations given by classical physics did not satisfy the observed reality. Bohr therefore presented his atomic model based on Max Planck's quantum theory and Einstein's photoelectric effect.

Before we look at the Bohr model itself, let's briefly discuss atomic spectroscopy.

1.5.1 Atomic spectroscopy

Atomic spectroscopy is an important area of research, both in the both in physics and chemistry. The emission spectra of chemical elements and compounds can be divided into three categories: continuous spectra, band spectra and line spectra, as stated by Tipler and Llewellyn (2010).

In the continuous spectrum, all wavelengths are found and are given by bodies in liquid or solid states, for which it is necessary to raise the body to high temperatures. Band spectra are bands formed by isolated molecules, not by atoms as in the line spectrum. Each band has a large number of very close lines that cannot be separated from each other.

Line spectra are characteristic of gaseous substances that contain only one type of chemical element, for example, the hydrogen atom. These spectra are inherent to each substance, i.e. the spectrum of a sample determines the substances that make up the material. Understanding the line spectra of hydrogen was a major step forward in Planck and Einstein's quantum theory, Id (2010).

An investigation of the spectra emitted by different types of atoms shows that each type has its own characteristic spectrum, i.e. a characteristic set of wavelengths at which the lines of the spectrum are found. This characteristic is of great practical importance because it makes spectroscopy another very useful technique to be added to the usual techniques of chemical analysis (EISBERG and RESNICK, 1973, p. 135).

When a hydrogen gas is subjected to an electrical discharge or heated to a high temperature, it emits light that passes through a diffraction network resulting in a spectral line, i.e. a set of lines with discrete wavelengths (RUSSELL, 1994), shown in Figure 11.

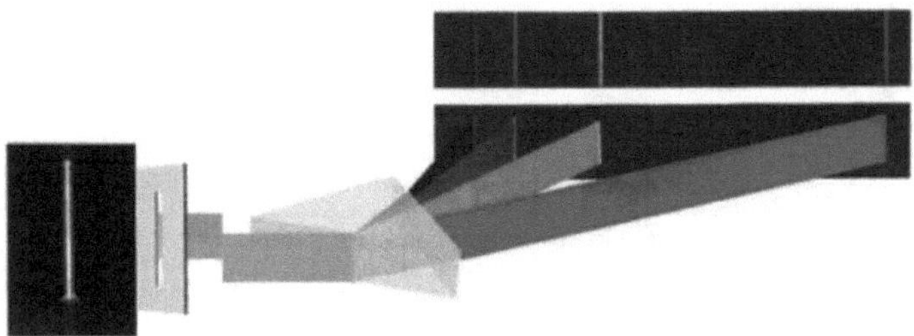

Figure 11 - Hydrogen atom spectrum.
Source: Positive note.

These lines have wavelengths in the region of visible light, and the regularities of the lines led the Swiss physicist Johann Jakob. Balmer obtained, in 1885, the empirical equation for wavelength, known as the Balmer series and given by:

$$\gamma = 364{,}6 \left(\frac{n^2}{n^2 - 4} \right), [nm] \tag{05}$$

where n are integers that take on values above 3. In 1890, Rydberg obtained for this expression

$$\frac{1}{\gamma} = R_H \left(\frac{1}{2^2} + \frac{1}{n^2} \right), \ n = 3, 4, 5,... \tag{06}$$

$$R_H = 1{,}097373 x 10^7 m^{-1}$$

where R_H is known as the Rydberg constant for hydrogen.

Based on this widespread knowledge, Niels Bohr proposed an explanation for these phenomena, introducing the ideas of Planck and Einstein

1.5.   2Bohr's postulates   for the hydrogen atom

In the first decade of the 1900s, as Halliday, Resnick and Walker (2011) show, scientists already knew that the hydrogen atom only emitted and absorbed four wavelengths of visible light. However, no scientist was able to explain why this phenomenon happened in this way. The explanations for these questions came from Bohr, who very boldly put aside some thoughts from classical physics and plunged into new ideas, since classical theory failed to explain the phenomena of the microscopic world.

In 1913 Bohr presented an atomic model that agreed with the experimental data for the hydrogen atom. His model was based on the following postulates, as shown by Eisberg and Resnick (1979):

- An electron in an atom moves in a circular orbit around the nucleus under the influence of the Coulombian attraction between the electron and the nucleus, obeying the laws of quantum mechanics;
- Instead of the infinity of possible orbits, which according to classical mechanics was continuous and allowed for infinite orbits, an electron can only move in an orbit in which its orbital angular momentum L is an integer multiple of ($\hbar = \hbar/2\pi$);
- Despite being constantly accelerated, an electron moving in one of these possible orbits does not emit electromagnetic radiation. Therefore its total energy E remains constant;
- Electromagnetic radiation is emitted/absorbed if an electron, which initially moves under an orbit of total energy E_i, changes its motion discontinuously so that it moves in an orbit of energy E_f. The frequency of the emitted radiation v is equal to the quantity ($E_i - E_f$) divided by Planck's constant h.

These postulates manage to completely blend classical and non-classical physics. The electron moving in a circular orbit is supposed to obey classical mechanics, yet the non-classical idea of quantisation of angular momentum is included. The electron is supposed to obey one feature of classical electromagnetic theory (Coulomb's law), yet it doesn't obey another feature (the emission of radiation by an accelerated charged body). However, we should not be surprised if the laws of classical physics, which are based on our experimentation with macroscopic systems, are not completely valid when we deal with microscopic systems such as the atom. (EISBERG and RESNICK, 1979, p 139)

1.5.3 Bohr's model

As mentioned earlier, in the classical model of the atom, the electron orbiting the nucleus in accelerated motion would radiate energy until it collided with the nucleus, as shown in Figure 12.

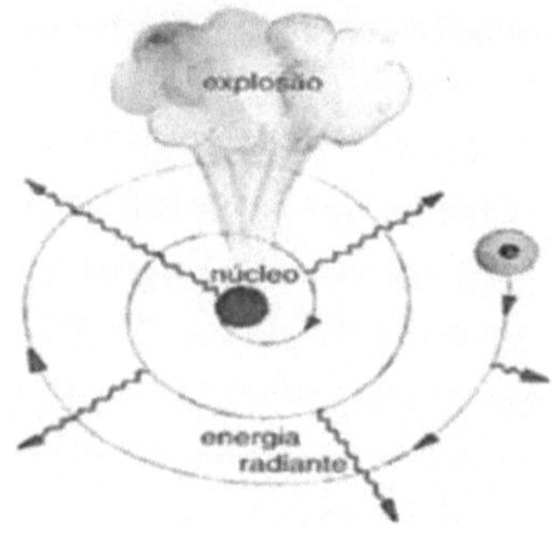

Figure 12 - Atomic model according to classical theory.
Source: Positive note.

In Figure 13 Bohr presents his model of the atom with the electron orbiting in a circle around the nucleus without emitting electromagnetic radiation according to his postulates.

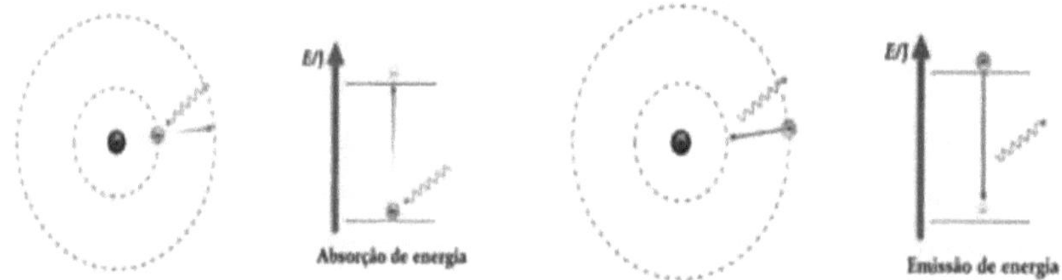

Figure 13 - Bohr's atomic model for the hydrogen atom.
Source: Positive note.

To this end, Bohr came up with an atom with a quantised orbital radius and energy and a discrete change of energy states, as presented by Halliday, Resnick and Walker (2011). According to the classical theory, for the orbital radius the force that keeps the electron in an orbit of radius *r is the* electrostatic force which, being a central force, we have that

$$F_{el} = F_c \tag{07}$$

where the electrostatic force is

$$F_{el} = k\frac{|q_1||q_2|}{r^2} \tag{08}$$

where **k** is the electrostatic constant, $k = 1/4\pi\varepsilon_0$, q_1 e q_2 **k** = 1/4TT£₀ , **q_r** and **q_s** are the charges of the proton and electron. And the centripetal force equals

$$F_c = m\frac{v^2}{r} \tag{09}$$

From the expression that equates the forces, applying the quantisation of angular momentum , = **mrv** = **nh,** Bohr finds the expression for the quantised radius for the deBohr mdel,

$$r = \frac{h^2 \varepsilon_0}{\pi m e^2} n^2, \text{ para } n = 1, 2, 3, \dots \tag{10}$$

can also be written as,

$$r = a n^2 \tag{11}$$

when $n = 1$ the electron is at the smallest radius allowed, known as the Bohr radius for the hydrogen atom, which is

$$a = \frac{h^2 \varepsilon_0}{\pi m e^2} = 5{,}291722 x 10^{-11} m \tag{12}$$

To demonstrate that the energy of the hydrogen atom is also quantised Bohr started from the fact that the total energy of the atom is equal to the sum of the kinetic energy K and the potential energy $U,$ as shown in equation (04), and by isolating mv^2 from expression (07) we arrive at the end of some calculations,

$$E = -\frac{1}{8\pi\varepsilon_0}\frac{e^2}{r} \tag{13}$$

Substituting expression (10) into (13) gives the quantised energy of the Bohr radius,

$$E = -\frac{me^4}{8\varepsilon_0^2 h^2}\frac{1}{n^2}, \text{ para } n = 1, 2, 3, \dots \tag{14}$$

Looking at this equation we can see that the quantisation of the orbital angular momentum has resulted in both the quantisation of the radius and the quantisation of the total energy of the electron, and furthermore, Eisberg and Resnick (1979) point out that the information in this equation can be shown in a diagram of energy levels. The diagram shows that the lowest energy value (highest absolute value) is for the first level with n equal to 1. When n increases, the energy increases (lowest absolute value). The diagram shows the energy of each level in electron volts on the right and the quantum number n of the corresponding level on the left.

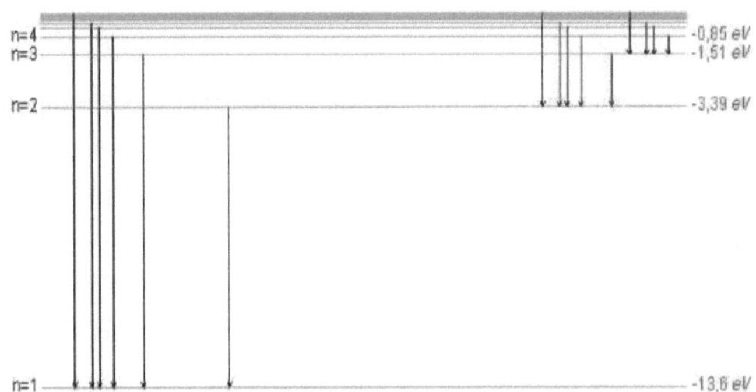

Figure 14 - Energy level diagram for the hydrogen atom (adapted)
Source: Physics Institute of the Federal University of Rio Grande do Sul.

When the electron changes its energy state, moving from one energy level Ei to another with energy En, *the* electron can emit or absorb energy. The frequency of the radiation it emits is discrete and given by

$$f = \frac{E_i - E_f}{h},$$ (15)

As we know, the frequency of an electromagnetic wave is given by the expression:

$$f = \frac{c}{\lambda}$$ (16)

where c is the speed of light and X is the wavelength of the electromagnetic wave. Applying (1) and (14) to (15) gives the Balmer series equation

$$\frac{1}{\gamma} = R_H \left(\frac{1}{n_f{}^2} - \frac{1}{n_i{}^2} \right)$$ (17)

which is the expression of the series of Lymanfn(n_f = 1), Balmer(n_f = 2), Paschen (n_f = 3), Brackett (n_f = 4), and Pfund (n_f = 5), which makes it possible to calculate the wavelength of the radiation emitted/absorbed by an electron when it changes energy level.

According to Eisberg and Resnick (1979, p. 139) **"the justification for** Bohr's postulates, or for any set of postulates, can only be found by comparing the predictions that can be obtained from the postulates **with experimental results". The authors also point out that according to the old** quantum theory, the energy states of an atom can only take on discrete energy values, since Bohr's model predicted that the total energy of the electron is quantised and Planck's theory of black body radiation also predicted that emission and absorption had quantised states.

Brennan (2003, p. 150) **reports that "one of Bohr's great achievements was to** show that it was not possible to describe the structure of the atom solely according to **classical** physics; **it was necessary to use quantum theory"**. According to the Group for the Re-elaboration of Physics Teaching - GREF (2007), the discoveries made by scientists were due to the conceptions of structures based on Bohr's proposal for the atom. The understanding of different electrical properties is based on Bohr's atomic representation of the atom.

The history of the atomic model has been compactly presented so far, from the first ideas about the constitution of matter to Bohr's model, which was a milestone in physics. From his model, together with the ideas of other physicists such as Planck and Einstein, the transition from classical physics to quantum physics began. Despite his sudden and short-lived success, his work contributed to the understanding of quantum physics and atomic structure. Bohr's model, explaining the behaviour of electrons when emitting and absorbing energy, made major contributions to new technologies through the study of semiconductor solids.

The success of Bohr's model was short-lived. And despite correctly predicting the wavelengths emitted and absorbed by the hydrogen atom, it was not possible to apply it to more complex atoms due to the fact that the electron behaves like a wave of matter confined in a potential well, as stated by Halliday, Resnick and Walker (2011).

Bohr's model was replaced by the new atomic model currently in force, which is based on the theories of Erwin Schrodinger, Louis Victor De Broglie and Werner Heisenberg. These new concepts became known as quantum mechanics, establishing the study of the behaviour of microscopic systems in terms of probabilities and no longer certainties. It was only possible to determine the region with the highest probability of finding an electron, represented by a wave function. (OLIVEIRA and GOMES, 2006)

According to Halliday, Resnick and Walker (2011), according to the Heisenberg Indeterminacy Principle proposed in 1927, it is not possible to simultaneously measure the position and momentum of a particle, as presented in Bohr's model, even with high-precision instruments. When the momentum is obtained accurately, the accuracy of the position is lost.

To obtain the correct energy levels for the electron, you need to solve the Schrodinger equation for a potential well represented by

$$U(r) = \frac{-e^2}{4\pi\varepsilon_0 r}. \tag{18}$$

Using the potential (18) in the three-dimensional, normalised Schrodinger equation, the result is the wave function of the ground state of the hydrogen atom, determined by the following expression

$$\psi(r) = \frac{1}{\sqrt{\pi}\, a^{3/2}} e^{-r/a} \tag{19}$$

Unlike other functions, this wave function has no physical meaning, only its square $(\psi^2(r))$ which can be interpreted as the probability per unit volume that an electron will be found in a given volume at a distance r from the centre of the atom. Considering the volume to be the region between two concentric shells and working with the radial probability density P(r), we arrive at the expression for the radial density of the fundamental state of the hydrogen atom:

$$P(r) = \frac{4}{a^3} r^2 e^{-2r/a} \tag{20}$$

Equation (20) simply shows that in a hydrogen atom the electron must be found somewhere, it only establishes the probability of it being found in that region, it does not give certainty of the exact position in which this electron is found.

Although Bohr's model didn't survive for long, it left great scientific contributions, such as semiconductors, the main subject of this work.

Semiconductors will be presented below, along with their very common technological applications in everyday life. Physics therefore has a very significant relevance in society and the study of this subject in the classroom should make clear this historical and contemporary relevance in the lives of students who see the subject as difficult and without application. As Bonadiman and Nonenmacher (2007) state, students arrive at secondary school with expectations of learning physics, but this almost always becomes a frustration that they will carry with them throughout their lives. He goes on to say that physics should be important to the student and that its teaching should have a significant relationship with the students' daily lives. In this way, we chose to talk about semiconductor materials, presenting their applications, which are quite intense in today's society.

CHAPTER 2

SEMICONDUCTOR MATERIALS

For Bogart (2001, p. 8) "semiconductors are the basis of many everyday electronic devices. Virtually all electronic devices are made from semiconductor materials". Semiconductors are neither perfect insulators nor perfect electrical conductors. Among the most widely used in industry are germanium and silicon, with silicon being the most widely used due to its large quantity on the face of the Earth, as stated by Peixoto (2001).

2.1 Atomic structure of semiconductors

According to Bogart (2001), in order to understand semiconductors, we must first understand the atomic structure of this material and its electrical properties. All atoms of an element have the same structure, and as we have already seen in atomic models, especially Bohr's, the atom is made up of a nucleus containing positive particles, the protons, and neutral particles, the neutrons, surrounded by negative particles, the electrons. The number of protons is equal to the number of electrons, thus making the fundamental atom electrically neutral.

The electrons in an atom are distributed in layers and every layer is divided into sub-layers according to the Pauli Exclusion Principle, which determines the exact number of electrons in each energy level.

The Pauli exclusion principle is a principle of quantum mechanics that applies to protons, electrons and neutrons. For electrons, it states that no two electrons with identical quantum numbers can exist in the same atom. "Between the quantum numbers n,l,mi and m_s of two electrons in the same atom there must be at **least one different number" (HALLIDAY;** RESNICK; WALKER, 2011, p 257).

The quantum numbers n,l,mi **and** ms are the principal, secondary, magnetic and spin quantum numbers respectively. The principal quantum number indicates the energy level or electronic layer in which the electron is located. The electronic layers are k, L, M, N, O, P and Q, represented by integers from 1 to 7. Secondary quantum number indicates the sublevel or subshell of the electronic layers in which the electron is located. The subshells are s, p, d, and f, and are represented by 0, 1, 2 and 3. The magnetic quantum number indicates the orientation of the orbitals in space, the region where there is the maximum probability of finding an electron in the atom. Each orbital holds a maximum of two electrons, which are represented by vertical arrows oriented upwards or downwards, depending on the spin quantum number. The spin quantum number is an intrinsic property of the electron and is associated with its angular momentum. Didactically, it indicates the rotation of the electron in the same orbital, where two electrons rotate in opposite directions, cancelling out each other's magnetism, thus guaranteeing a stable atomic system. The spin values for the electron are represented by +1/2 and -1/2. By convention, the direction of the arrow, up or down, indicates the electron's spin (RUSSEL, 1994).

All electrons are denoted by their four quantum numbers, which indicate their distance from the nucleus, the shape of their orbit, their magnetic moment and their spin. According to the Pauli exclusion principle, there can only be two electrons in the same energy level and they must have different spins (GREF, 2007).

The electron capacity in each layer, shown in Figure 15, is 2 electrons in the K layer, 8 in the L layer, 18 in the M layer, 32 in the N layer, 32 in the O layer, 18 in the P layer and 4 in the Q layer, respecting the Pauli exclusion principle. The electrons in the layers are divided into sub-layers, 2 in the s, 6 in the p, 10 in the d, and 14 in the f, respectively (BOGART, 2001; GREF, 2007).

Figure 15 - Representation of the electronic layers of an atom.
Source: Just chemistry.

Let's look at the example of silicon in Figure 16, the most common semiconductor material.

used in the manufacture of electronic devices. Silicon makes up 27.7 per cent of the earth's crust, second only to oxygen, but is only found in its combined form with other elements. It is found in practically all rocks, sands, clays and soils. It can also be found in natural waters, dust, plants, skeletons and the organic fluids of some animals. Its natural compounds are quartz, asbestos, zeolite and mica. These silicon compounds can be used in the manufacture of bricks, different concretes, glass, enamels, varnishes, ceramics, oils, grease, rubbers, radar, lighters and watches. It has important uses in plastic surgery, prosthetics and construction. When the aim is to obtain ultra-pure silicon for the manufacture of semiconductors, it is prepared by decomposing silane (SiH_4) or silicon tetrahalides at high temperatures. Due to its electronic structure, silicon is very important in the electronics industry as a semiconductor. Semiconductors are the basic materials used in the construction of computer chips, transistors, silicon diodes and many other electronic devices (PEIXOTO, 2001).

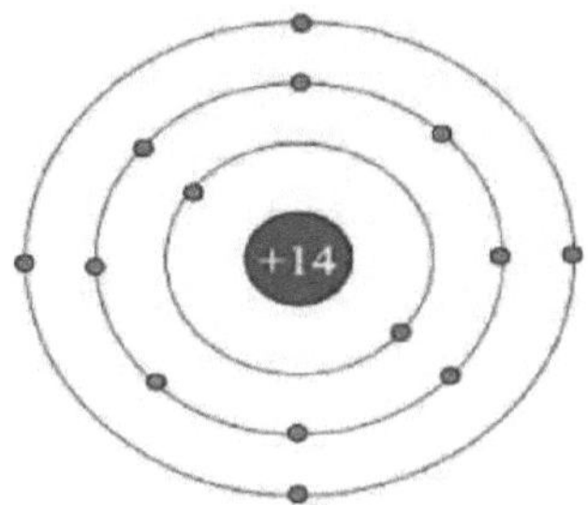

Figure 16 - Electron distribution of the silicon atom with four electrons in the last layer.
Source: Guaratinguetá Engineering Faculty of the São Paulo State University - FEG-UNESP.

The electron distributions of a chemical element obey the predicted solutions of the Schrodinger equation and the Pauli exclusion principle. In the case of silicon, we have 14 electrons orbiting the nucleus with 14 protons and 14 neutrons. There are 2 electrons in the first layer, 8 in the second layer and 4 in the third layer.

The last energy layer containing electrons for an atom in the ground state is called the valence layer. In this layer, the silicon atom has exactly four electrons, a characteristic of semiconductor materials for the valence layer.

The atoms of these materials form a monocrystalline structure, i.e. a firm and stable bond with interlaced structures, (BOGART, 2001). In a crystalline structure, whether silicon or germanium, each atom containing four electrons in the valence shell will share its electrons with four neighbouring atoms, thus filling its p subshell. The sharing of electrons in an atomic bond, Figure 17, is called a covalent bond.

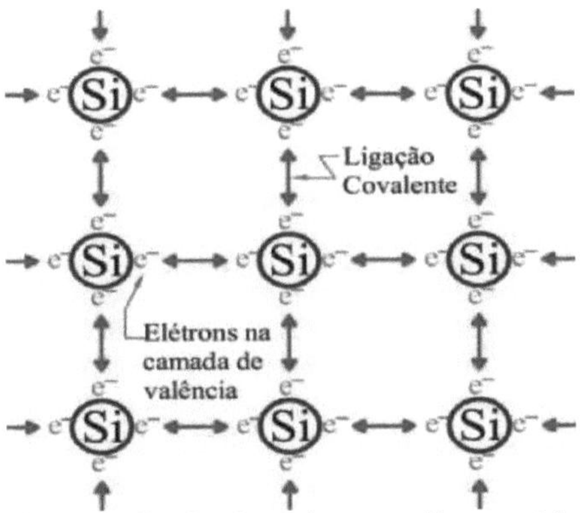

Figure 17 - Covalent bonds in a pure silicon crystal.
Source: State University of Western Paraná - UNIOESTE.

2.   2 Electric current   in semiconductors

As mentioned earlier, semiconductors are intermediate materials between insulators and conductors. The difference between them can be seen in the energy band diagrams in Figure 18. An energy band is the set of energy levels that electrons have in a solid. In an isolated atom, the electrons have defined energy levels and in a solid, due to the large number of closely packed atoms forming a crystalline lattice, the energy levels overlap and become energy bands (EISBERG and RESNICK, 1979).

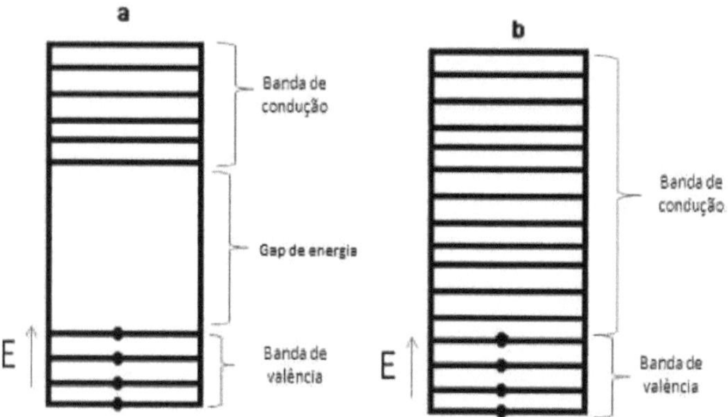

Figura 18   - a) distribution of energy levels of insulating material. b) distribution of energy levels of conductive material. (adapted)
Source: GREF.

The width of the forbidden band, i.e. the intermediate energy level that the electron cannot assume, is called the energy gap. The energy gap is the minimum energy that the electron must acquire to pass from the valence band to the conduction band or the energy it must release to return to the valence layer. Insulating materials are those where the distance between the valence band and the conduction band is very large, i.e. the energy gap. Conductors are materials where the valence and conduction bands overlap, or where the energy gap is very small or zero (GREF, 2007).

Considering the definition of a semiconductor and Figure 19,

> "The use of energy diagrams makes it possible to interpret a third group of substances whose separation between the band already occupied by electrons and the free band is smaller when compared to insulators. Thus, with a certain amount of energy, such as a photon of light or thermal energy, the electrons reach the free band, making the substance conductive. This is why such substances are called semiconductors" (GREF, 2007, p. 282).

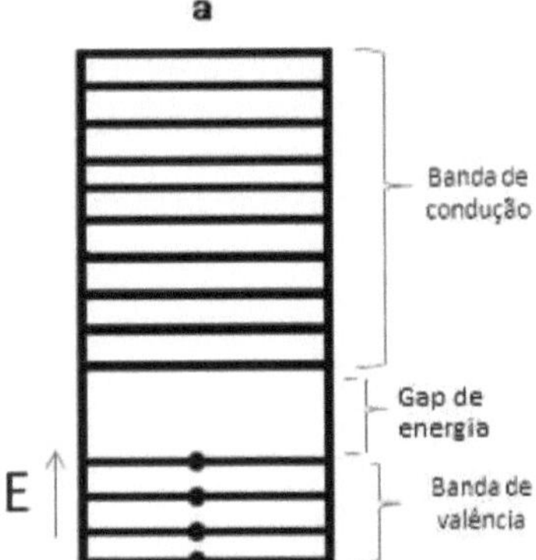

Figura 19   - Energy diagram for semiconductors (adapted)
Source: GREF.

According to Bogart (2001, p. 9-10):

"Electrons are in the conduction band because they are charge carriers that contribute to the conduction of current. Electrons in the valence band have less energy and are shown at the bottom of the energy diagram. The region between the conduction and valence bands is called the forbidden band because, according to quantum theory, it is not possible for electrons to accumulate energy and remain in these energy levels."

For semiconductors, such as silicon and germanium, the energy gaps at room temperature are 1.14 eV and 1.67 eV, respectively, while in insulators, such as diamond, this value is 5.5 eV, while in conductors there is no such energy difference because the bands overlap (HALLIDAY; RESNICK; WALKER, 2011). According to Eisberg and Resnick (1979, p. 587), "a semiconductor is a covalent solid that can be considered an insulator because its valence band is full and the conduction band is completely empty at absolute zero [...]". Bogart (2001, p.10) tells us that

"As the temperature rises, the electrons accumulate enough energy to cross the forbidden band and reach the conduction band. For a semiconductor, an increase in temperature [...] [the resistance decreases], meaning that a semiconductor has a resistance with a negative temperature coefficient."

In conductors, the increase in temperature causes the number of electrons in the conduction band to increase, but this increase is much greater than in semiconductors, which causes an increase in collisions between electrons, making electrical conduction more difficult. The result is that it is difficult for charges to flow in one direction through the conductor at high temperature, making its resistance have a positive temperature coefficient.

In semiconductors, the effect of increasing temperature causes the number of carriers to increase significantly due to the small energy gap. When the electrons in a semiconductor receive enough energy to overcome the energy gap barrier, these electrons jump to the conduction layer leaving a gap in the valence layer, which function as positive charge carriers. As we can see in Figure 20.

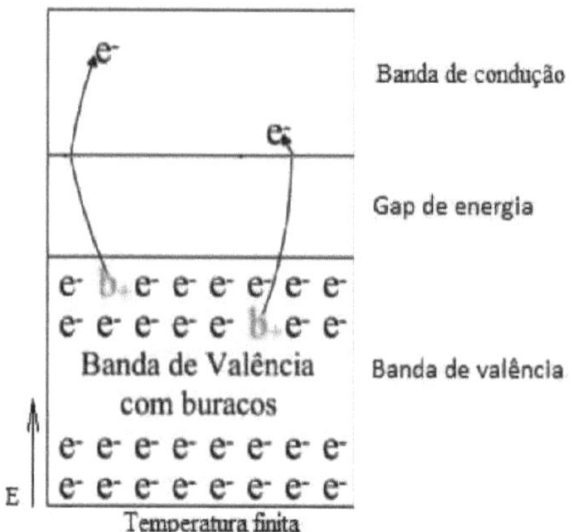

Figure 20 - Electrons excited to the conduction layer (adapted)
Source: UNIOESTE.

According to Eisberg and Resnick (1979), holes (gaps) left by electrons in the valence band will be occupied by other electrons contributing to the conductivity of the material. "Moving to energy levels in the conduction band, the electron leaves a vacant place in the valence band, called an electronic gap, which could be interpreted as the emergence of a local positive charge." (GREF, 2007, p. 283).

In semiconductors, as well as the movement of electrons into the conduction band, there is also the movement of electrons towards empty spaces, gaps, which consequently leave other gaps to be occupied. Both the electrons that move into the conduction band and those that occupy the gaps play a part in electrical conduction in semiconductors. In this way, electrons and gaps form electrical conduction in semiconductors, in quantities of

equal charges in the case of pure substances, (GREF, 2007). According to Bogart (2001, p. 11) "The gap current occurs at the level of the valence band, because the electrons in the valence band do not become free electrons when they simply move from one atom to another".

The same author also emphasises that pure semiconductor material is called intrinsic. The interesting thing about pure semiconductor materials is the possibility of altering their electrical conductivity by adding impurities. This process is known as doping (TIPLER and LLEWELLYN, 2011).

2.3 N-type and p-type doped semiconductors

"The versatility of semiconductors can be greatly increased by introducing a small number of atoms (called impurities) into the crystal lattice; this process is called doping." (HALLIDAY; RESNICK; WALKER, 2011, p. 289). Doping can be done in two ways: by increasing the number of positive charge carriers, known as p-type doping, and by increasing the number of negative charge carriers, known as n-type doping. As we saw earlier, semiconductors conduct energy through both electrons and gaps. Doping will precisely increase one of these charge carriers. Doping a semiconductor such as silicon and germanium basically consists of putting a small concentration of another substance into the silicon or germanium.

2.3.1 Type n doping

The silicon atom contains four electrons in the valence layer and makes covalent bonds in a crystal lattice with other atoms of the same substance. If a silicon atom is removed and an atom containing five electrons in the valence layer, i.e. the last layer, is placed in its place, such as arsenic, phosphorus, etc., we say that the semiconductor has been doped with an n-type impurity, because four electrons from this last atom will make covalent bonds with the silicon atoms, leaving an extra electron weakly attracted to the nucleus, as shown in Figure 21 for doping with the arsenic atom,

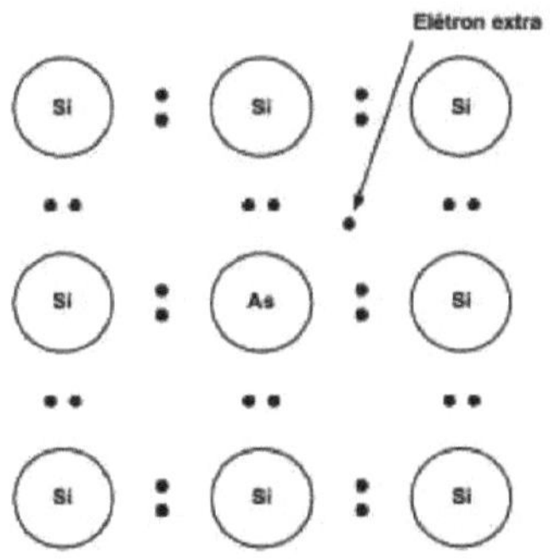

Figure 21 - Silicon crystal doped with arsenic, containing five electrons in the valence layer, leaving one extra electron.

Source: Materials Engineering Department of the University of São Paulo - DEMARUSP.

According to Halliday, Resnick and Walker (2011), in an intrinsic silicon covalent bond, the electrons are in the valence band of the semiconductor, moving to the conduction band only when they receive enough energy to be pulled out of the covalent bond. However, in an n-type doped semiconductor, the extra electron will occupy an energy level very close to the conduction band, even before a thermal agitation or potential difference, and the energy required to transfer this electron to the conduction band is small.

The atom inserted into the crystal lattice is called a donor, because it donates electrons to the conduction band. Donor atoms contain five electrons in the last layer and only four will participate in the covalent bond with the silicon atoms. The term n-type semiconductor comes from the fact that electrons are the majority charge carriers. With thermal agitation, some of the electrons in the valence band, along with the extra electrons, move to the conduction band. When the electron leaves the valence band, it leaves a gap, a hole (Figure 22). In the conduction band there will be extra electrons from the atom inserted in the silicon plus electrons from the valence band, which is why in n-type semiconductors, electrons are the majority carriers and gaps are the minority carriers (HALLIDAY; RESNICK; WALKER, 2011; TIPLER and LLEWELLYN, 2010).

Figure 22 - Extra electrons from the donor atom occupying a level close to the conduction band (adapted).
Source: UNIOESTE.

2.3.   2P-type semiconductors

Similar to what we saw in the n-type case, the material must be similar in size to the silicon atom. In p-type doping, atoms with three electrons in the valence layer are inserted into the intrinsic silicon crystal lattice, for example aluminium, indium, boron, etc. The atom inserted into the silicon crystal, boron, shown in Figure 23, will make covalent bonds with the silicon atoms. Boron has only three electrons available in the valence layer to share with the crystal lattice, leaving an empty space that can receive an electron from the lattice, causing the position of empty spaces to vary, thus configuring a movement of gaps or holes. As the atom has one less electron and can accept another, it is called an acceptor.

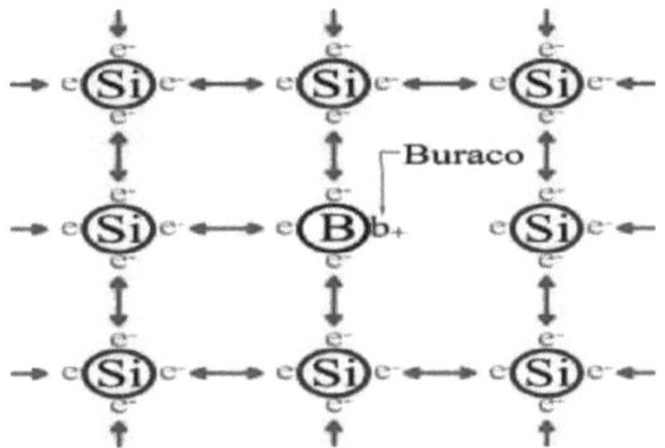

Figure 23 - Silicon crystal doped with an atom containing three electrons in the valence layer, leaving an acceptor hole.
Source: UNIOESTE.

The gaps, holes, will occupy an energy level between the

valence band and the conduction band, but close to the upper limit of the valence band. As shown in Figure 24, the electrons in the valence band jump to the level slightly above this band, leaving empty spaces in the valence band. The energy required for this transfer is much lower than the energy required for transfer to the conduction band.

Semiconductors doped with acceptor atoms are called p-type semiconductors, to indicate positive charge carriers, holes, in the valence band. In p-type semiconductors, holes are majority carriers and electrons are minority carriers (HALLIDAY; RESNICK; WALKER, 2011; TIPLER and LLEWELLYN, 2010).

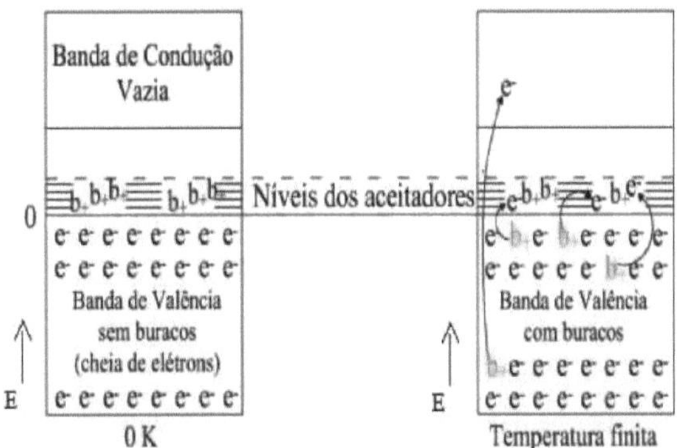

Figure 24 - holes in acceptor atoms occupying an energy level close to the valence band (adapted).
Source: UNIOESTE.

## 2.	4P-N junction	, contact effect between regions

If an n-type semiconductor and a p-type semiconductor are placed in contact, as shown in Figure 25, and their temperature is varied, a p-n junction is formed. "A p-n junction [...] is a semiconductor crystal that has been doped in one region with a donor impurity and in the other region

with an acceptor impurity. This type of junction is present in practically all semiconductor devices."
(HALLIDAY; RESNICK; WALKER, 2011, p. 292).

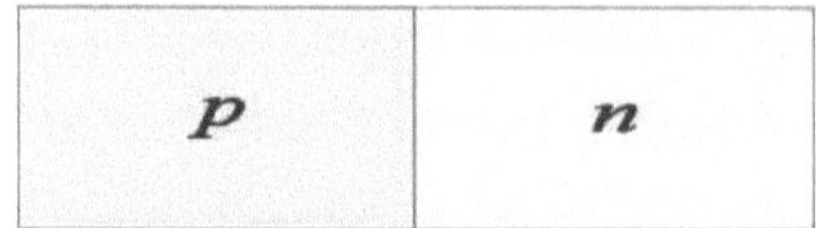

Figure 25 - p-n junction.
Source: FEG-UNESP.

Upon contact, Figure 26, and temperature variation, charge carriers in the majority of both the p and n regions begin to diffuse, the gaps from the p region to the n region, and the electrons from the n region to the p. These movements of charge carriers give rise to a diffusion current, (Idif). The movement of electrons from the n-side to the p-side leads to a concentration of positive charges at the end of the junction on the n-side. When these electrons combine with the gaps on the p-side, a concentration of negative charges arises at the end of this side (GREF, 2007; HALLIDAY; RESNICK; WALKER, 2011).

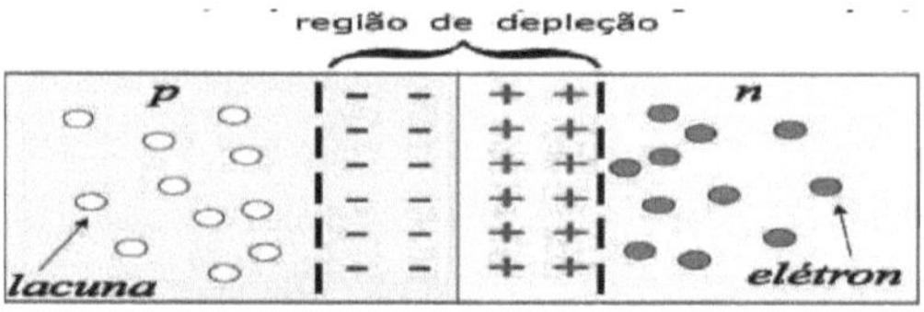

Figure 26 - P-N junction with depletion region formation.
Source: FEG-UNESP.

This process results in the formation of a space charge on both sides, stopping the movement of the majority of charge carriers and giving rise to a depletion zone. At the ends of the depletion zone, due to the formation of the space charge, there is a difference in contact potential that prevents charge carriers from migrating from one region to another (HALLIDAY; RESNICK; WALKER, 2011).

The majority carriers give rise to the diffusion current, but we also have another current due to the movement of the minority charge carriers. On the n side of the junction, we have gaps as minority carriers and on the p side, electrons. The potential difference V0 is a barrier that prevents the majority carriers from moving, but facilitates the minority carriers, p or n, providing the drift current (Ider). The direction of the drift current (Ider) in a p-n junction is from right to left, the opposite direction to the diffusion current (Idif). In an unpolarised p-n junction, the total current is zero because the

diffusion and drift are in opposite directions.

The great importance of semiconductors lies in the possibility of manufacturing devices to be used in electronic equipment. The rectifier diode, the light-emitting diode (LED) and the transistor are semiconductor devices that are widely used in electronic devices that are part of people's daily lives.

35

CHAPTER 3

SEMICONDUCTOR APPLICATIONS

Teachers should frequently learn about new technologies and discuss this information in the classroom, as technological evolution is present in the students' social environment. According to Ferreira, Bampi and Alvares (2004), semiconductors are the driving force behind technological development, making it possible to advance information technology, communications and their applications, thus revolutionising society's day-to-day life.

The teacher must be prepared to pass on this information in the classroom, since the content taught must make sense to the daily lives of both the teacher and the student. Oliveira and Hosoume (2008) emphasise the importance of teachers knowing physics and understanding the existence of a scientific construct that has enabled significant advances for society. According to the National Curriculum Parameters - PCN (2000):

> "Physics is knowledge that allows us to develop models of cosmic evolution, to investigate the mysteries of the sub-microscopic world, of the particles that make up matter, while at the same time allowing us to develop new sources of energy and create new materials, products and technologies. Incorporated into culture and integrated as a technological tool, this knowledge has become indispensable to the formation of contemporary citizenship. It is hoped that the teaching of physics in secondary schools will contribute to the formation of an effective scientific culture that allows individuals to interpret natural facts, phenomena and processes, situating and dimensioning the interaction between human beings and nature as part of nature itself in transformation. To this end, it is essential that physical knowledge is explained as a historical process, the object of continuous transformation and associated with other forms of human expression and production. It is also necessary for this culture in Physics to include an understanding of the set of equipment and procedures, technical or technological, of everyday domestic, social and professional life."

Knowing the importance of semiconductors in technological advances, here are some applications of this material that is present in society.

3.1 Rectifier diode

The rectifier diode, shown in Figure 27, is a semiconductor device used mainly to transform alternating voltage into direct voltage. Considering the alternating mains voltage, the rectifier diode has a very important application.

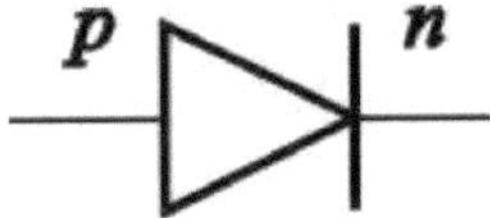

Figure 27 - Representation of the diode.
Source: FEG-UNESP.

In a circuit with a p-n junction semiconductor, the electric field produced by the energy source and the electric field of the p-n junction are superimposed. If the direction of the fields are opposite, charges will move through the semiconductor device; however, if the fields are the same, the existing barrier in the p-n junction (depletion zone) will increase, making it difficult for electric current to pass through. In this way, the semiconductor will allow current to pass in one direction, direct polarisation, and practically retain current in the opposite direction, reverse polarisation (GREF, 2007).

In direct polarisation (Figure 28), the p-side of the junction becomes "more positive" by reducing the number of negative charge carriers in this region, and the n-side becomes "more negative" by reducing the number of positive charge carriers. The result of this effect produced by direct polarisation is a decrease in the depletion zone, thus restoring the diffusion of the majority of charge carriers. This means that the diffusion current becomes greater than the drift current, resulting in a direct current (HALLIDAY; RESNICK; WALKER, 2011).

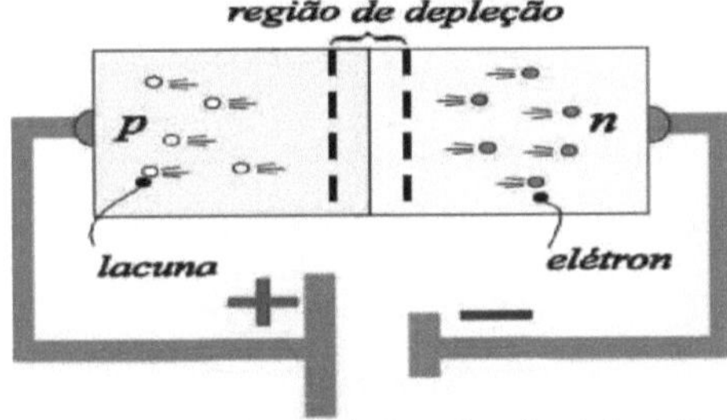

Figure 28 - Directly polarised p-n junction (adapted)
Source: FEG-UNESP.

In reverse polarisation, Figure 29, the free electrons in the n region move away from the junction towards the positive pole of the voltage source and the gaps on the p side towards the negative pole of the source. As electrons and gaps move away from the junction, there is a greater concentration of negative charges on the p-side of the junction and positive charges on the n-side. This movement causes the depletion zone (region) to increase, causing a decrease in the diffusion current and a slight increase in the drift current, resulting in a small reverse current.

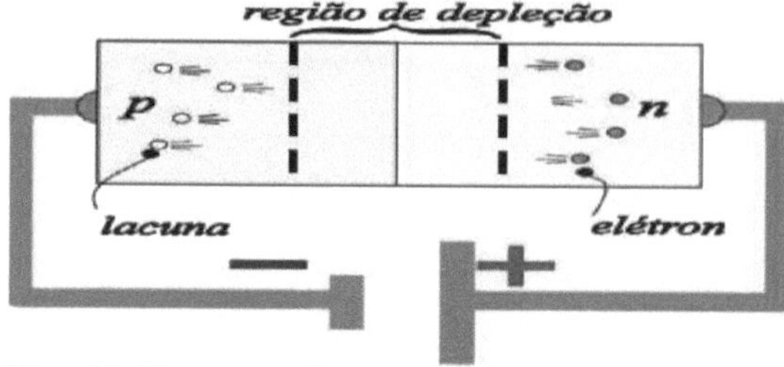

Figure 29 - Reverse-polarised p-n junction (adapted)
Source: FEG-UNESP.

3.   2 Light Emitting Diode (LED)

When electrons in the conduction band occupy gaps in the valence band, the energy difference, which is equal to the energy gap, is released in some way. In silicon and germanium, this energy is emitted in the form of vibrations in the crystal lattice, but in others such as gallium arsenide (GaAs) or gallium phosphide (GaP), this energy can be emitted in the form of a photon of light, i.e. electromagnetic radiation (GREF, 2007; HALLIDAY; RESNICK; WALKER, 2011).

In a rectifier diode, energy is released in the form of vibrations in the crystalline network, unlike LEDs (Light Emitting Diode), Figure 30, which are manufactured to release energy in the form of light and their physical structure allows them to concentrate light in a specific direction. The great advantage of LEDs over conventional incandescent bulbs is the absence of a filament, which guarantees greater durability, as well as their size, which facilitates their insertion into much smaller electronic circuits (UNIVERSIDADE FEDERAL DE UBERLÂNDIA, 2009).

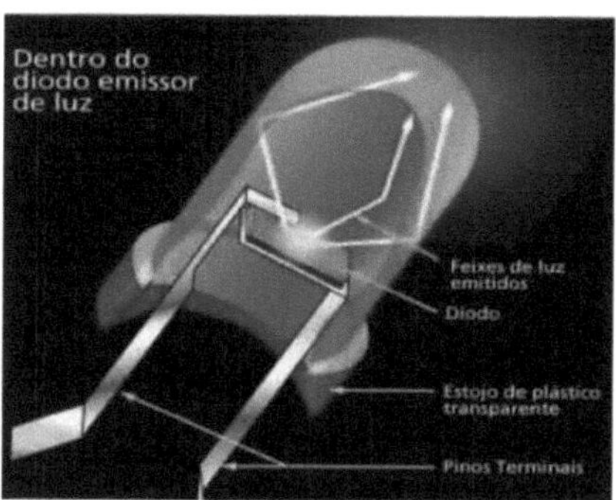

Figure 30 - Light-emitting diode.
Source: Federal University of Uberlândia.

LEDs are used in televisions to form images and illuminate screens, in digital clocks to display the number and illuminate it, to transmit information in remote controls, microwave ovens, computers, radios, mobile phones, in short, in a multitude of electronic devices.

3. 3 Organic Light Emitting Diode (OLED)

The OLED (Organic Light-Emitting diode) basically has the same structure as LEDs, but differs in the type of material used, with electroluminescent substances composed of carbon that, when excited by an electric current, emit light at a frequency determined by their chemical composition. In other words, they are diode cells printed on the screen that are polarised according to the image. This technique makes it possible to build very small or large monitors that are water-resistant due to their plastic nature, and flexible or even foldable, as we can see in Figure 31 the structure of an OLED and Figure 32, the applications of this technology in flexible screens. (UNIVERSIDADE FEDERAL DE UBERLÂNDIA, 2009).

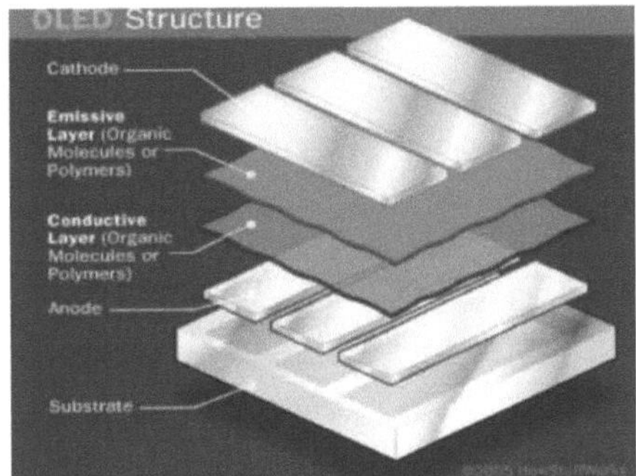

Figure 31 - OLED structure.
Source: Federal University of Uberlândia.

Figure 32 - OLED technologies in microfilm screens.
Source: Federal University of Uberlândia.

3.4 Bipolar Junction Transistor

In 1947, John Bardeen and Walter Brattain, supervised by William Shockley at Beel Telephone Laboratories in the United States, built the **transistor, which means "transfer resistor".** The discovery of this device was accidental, occurring while they were studying the contact point diode. Despite being a major scientific breakthrough, the transistor was not immediately successful due to manufacturing difficulties and the high price of germanium, a rare element. Early transistors were made of germanium, but it was soon discovered that silicon offered a number of advantages over germanium. In 1955, the first silicon transistor was commercialised (PONTIFÍCIA

UNIVERSIDADE CATÓLICA DO RIO GRANDE DO SUL, 2006).

The transistor comes from a combination of p- and n-type semiconductors, and can have a pnp or npn configuration. It replaces triode valves in amplifiers and other electronic circuits. Unlike the valve, the transistor consumes no energy and is made in microscopic proportions, making hundreds of transistors fit on a chip just a few millimetres in size. The transistor amplifies and switches signals, replacing thermionic valves in new technologies, making it an important component in electronics. It is a device with three terminals, as shown in Figure 33 and the device in Figure 34, and it is possible to use the voltage between two of the terminals to control the flow of current in the third terminal. (UNIVERSIDADE FEDERAL DE UBERLÂNDIA, 2009).

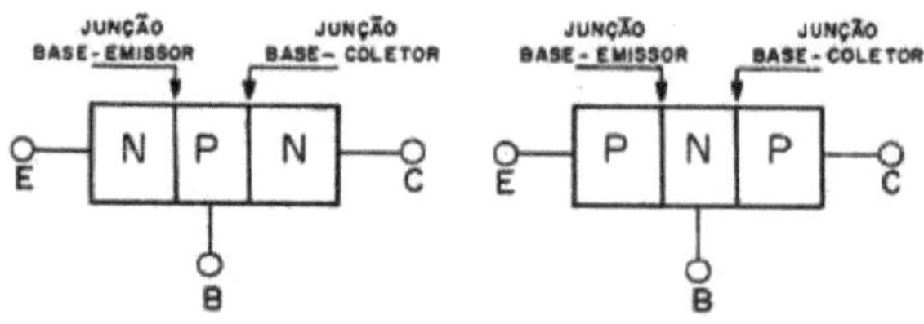

Figure 33 - Representation of the bipolar junction.
Source: Radio Amateurs website.

Figure 34 - Bipolar junction transistor.
Source: Maintenance and Supplies.

Processors were the biggest beneficiaries of the invention of transistors, being made up of billions of transistors connected together. The transistor is the basis of modern computing, as its creation has given it a breakthrough that would not have been possible without transistors. It is responsible for controlling the flow of data in various pieces of equipment, with the processor being the main user of this technology (TECMUNDO, 2010).

3.   5Thermistors

Thermistors, Figure 35, are semiconductor resistors that are very useful in electronic devices and are sensitive to temperature. They are classified into two types according to their specificity: PTC (Positive Temperature Coefficient) and NTC (Negative Temperature Coefficient) thermistors. PTC thermistors have applications such as temperature measurement, control and compensation and overheating protection, as well as applications such as overload protection, liquid level detection, air flow detection and constant temperature. NTC thermistors are used as temperature probes in medical devices, household appliances, the automotive sector, telecommunications and others. In some cases, they measure absolute temperature values and in others they measure with precision of up to one hundredth of a degree, (MECATRÔNICA ATUAL, 2013; UNIVERSIDADE FEDERAL DO PARANÁ, 2000).

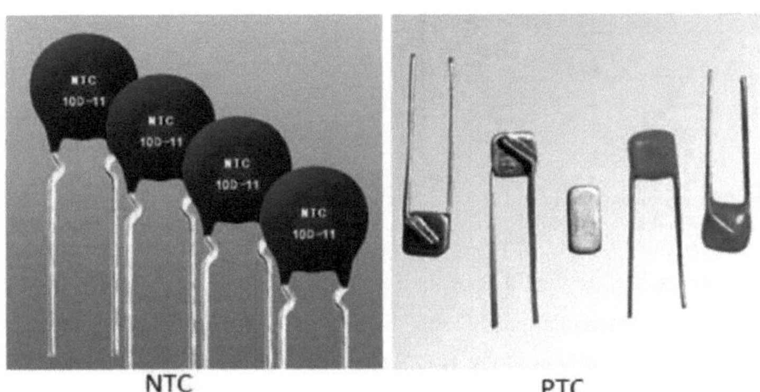

Figure 35 - NTC and PTC thermistors.
Source: Engineers garage.

Air conditioners, refrigerators and freezers, dehumidifiers, hydromassage heaters, therapeutic equipment, incubators, coffee makers, ovens and autoclaves, fryers, showers, dental equipment, water filters, incubators, electronic thermostats, fast food machines, thermal gondolas, displays, energy management, are examples of everyday equipment that makes use of this semiconductor device. (UNIVERSIDADE FEDERAL DE UBERLÂNDIA, 2009).

3. 6Photoconductors

The photoconductor is a p-n junction semiconductor device that is sensitive to electromagnetic radiation and its conductivity varies with the incidence of light, the more light the greater the electrical conduction capacity, as shown in Figure 36.

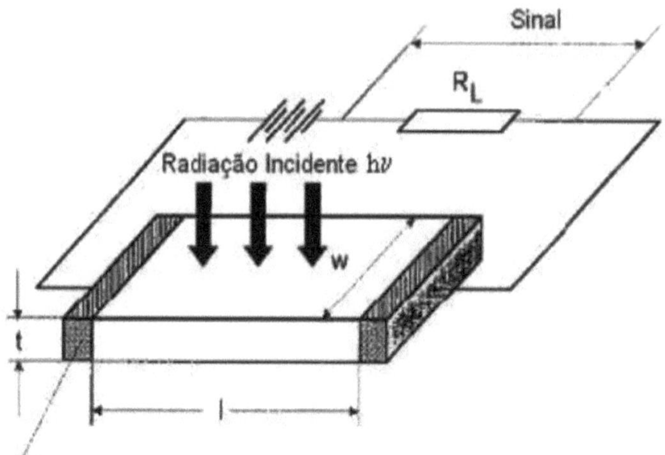

Figure 36 - Representation of the photoconductor.
Source: Federal University of Uberlândia.

A photon with energy greater than that of the forbidden band gap is absorbed by the photoconductor, producing an electron-lacunar pair, and the light energy is converted into electrical energy. An example of a photoconductor is the LDR (Light Dependent Resistor). The fact that its electrical resistance decreases makes it possible to use this component to develop sensors that are activated or deactivated when they are affected by light energy.

The LDR can be used to automatically control doors, security alarms against unauthorised entry and street lighting, all of which are photo-controlled for relay operation. Commercial photoconductive devices are called photoconductive cells and are used to measure the amount of illumination, and as an on-off light relay, in this case called a photocell relay, illustrated in Figure 37. It is able to perceive sunlight, thus identifying whether it is day or night, switching on the lamps automatically when it gets dark and switching them off after dawn, with great application in public lighting. Photoconductors are also used in xerox machines. At the start of the operation, a lamp is switched on, which scans the entire document to be copied, then the image is projected onto the surface of a photosensitive cylinder (made of aluminium, coated with photoconductive material). (UNIVERSIDADE FEDERAL DE UBERLÂNDIA, 2009).

Figure 37 - Photoelectric relay.
Source: Federal University of Uberlândia.

3.   7Photovoltaic cells

Figure 38 shows a representation of a solar cell structure that captures solar radiation and transforms it into electrical energy.

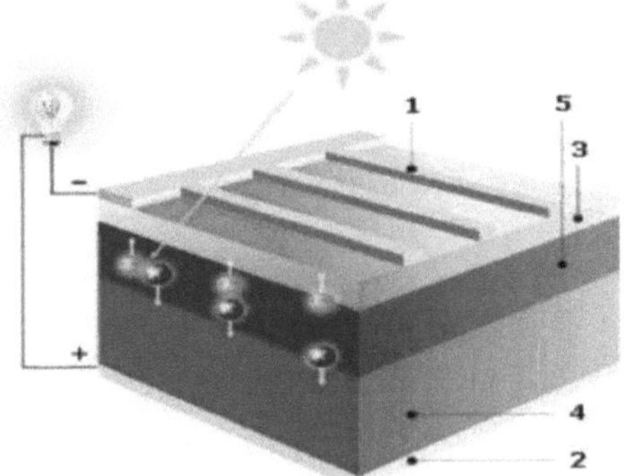

Figure 38 - Representation of a photovoltaic cell. 1) negative electrode; 2) positive electrode; 3) n-type layer; 4) p-type layer; 5) boundary layer (depletion).
Source: Federal University of Uberlândia.

> When the electrons and gaps reach the PN junction, they are separated by the internal field of the depletion region. The photovoltaic element forces current to flow in the external circuit, so light energy is converted into electrical energy. In other words, the solar cell works on the principle that the incident photons, colliding with the atoms of certain materials, cause the negatively charged electrons to move, generating an electric current. (UNIVERSIDADE FEDERAL DE UBERLÂNDIA, 2009, p. 94).

The set of photovoltaic cells are called solar panels, like the ones in Figure 39.

Figure 39 - Solar panels.
Source: Solar energy.

The electrons in some materials exhibit the photovoltaic effect when in the presence of light. These materials include silicon and gallium arsenide. For this effect to occur, it doesn't need the brightness of the sun, as the photovoltaic effect also occurs with a low intensity of sunlight.

Due to the reflection of sunlight, days with few clouds can result in more energy production than completely clear days. Currently, the Ministry of Mines and Energy is developing various projects to harness solar energy in Brazil, particularly through photovoltaic systems for generating electricity, with the aim of serving rural communities and/or those isolated from the electricity grid and regional development (BRASIL).

CHAPTER 4

FINAL CONSIDERATIONS

The work presents the evolution of the idea of the atom up to Bohr and the study of semiconductors with some applications, precisely to show that Physics is inserted in everyday life. Semiconductors were chosen because they are the main driving force behind technological advances, as stated by Ferreira, Bampi and Alvares (2004).

Knowledge of physics has evolved and provided improvements in human life, as shown in this work with the conceptions of the atom and technological applications of semiconductors, this knowledge has brought great benefits to the present day.

However, the knowledge produced by humanity taught in the classroom must be dynamic and not boring, and as Gadotti (2003, p. 47) states, "The teacher needs to know, however, that it is difficult for the student to realise this relationship between what they are learning and the legacy of humanity". From my experience as a high school student, I couldn't see the point or application of physics in the classroom. I often asked and heard questions from my classmates: why study physics? why do I have to study physics?

Physics has applications in people's lives, but often these applications are not shown in the classroom. The worst answer to the above question would be because the student needs to pass. Freire (1996) states that a good teacher is one who engages the student, manages to hold their attention on the content being taught, and also emphasises that the teacher has to be curious about learning, getting to know and awakening this curiosity in the students.

By knowing the physics behind the technologies, not just this physics, it is possible to answer the questions asked by the students and give meaning to the content, to show them that physics has applications and meaning in the life of humanity, to show that they make use of the physics taught in the classroom every day, whether using a mobile phone, the television, taking a bath in a hot shower, making use of electro-electronic equipment, to show that this equipment has evolved substantially thanks to the evolution of physics.

As a result, the teacher must relate the content to the student's everyday life, arousing curiosity in the student to understand the phenomena observed in everyday life. Gadotti (2003) says that students only want to learn when what they are taught makes sense to their reality, otherwise they won't want to learn.

When the student seeks to make sense of the study of physics, with questions or in any other way, the teacher has the option of answering them by talking about new technologies, addressing their principles of operation, explaining why devices are smaller than those of yesteryear and use less energy. By understanding semiconductor electronics, the teacher will be able to arouse the student's curiosity to understand physical phenomena. However, in order to do this, the teacher must seek out this knowledge and pass it on, because as Freire (1996, p. 37) states, "I cannot teach what I do not know", which means that the teacher must prepare themselves as much as possible so that, when questioned, they don't lie or give any answer, or don't always have to say that they don't know.

CHAPTER 5

REFERENCES

BIBLE. Portuguese. **Holy Bible.** João Ferreira de Almeida. Barueri: Bible Society of Brazil, 2003. 1536 p.

BOGART Jr., Theodore. The theory of semiconductors. In **Devices and electronic circuits.** v. 1. 3 ed. São Paulo: Pearson, 2001, chap 1, p. 5-58.

BONADIMAN, Helio; NONENMACHER, Sandra Elisabet B. **Enjoying and Learning in Physics Teaching:** a methodological proposal. Caderno Brasileiro de Ensino de Física, v. 24, p. 194-223, 2007.

BRAZIL. Ministry of Education. **National curriculum parameters:** secondary education. Available at: < http://portal.mec.gov.br/seb/arquivos/pdf/ciencian.pdf> accessed on: 20 June 2014.

BRAZIL. Ministry of the Environment. **Solar energy.** Available at <http://www.mma.gov.br/clima/energia/energias-renovaveis/energia-solar> accessed on: 19 June 2014.

BRENNAN, Richard P. **Giants of physics:** a history of physics through eight biographies. Translated by Maria Luiza X. de A. Borges. Rio de Janeiro: Jorge Zahar, 2003.

FORTALEZA MILITARY SCHOOL. **Rutherford's atomic model.** Available at :< http://www.cmf.ensino.eb.br/sistemas/pipa/files/arquivos/NA/1952_08_05_2013_Arq uivo.pdf> accessed on: 08 May 2014.

DEPARTMENT OF MATERIALS ENGINEERING, UNIVERSITY OF SÃO PAULO. **Semiconductor materials.** Available at: <http://www.demar.eel.usp.br/eletronica/aulas/Semicondutores.pdf> accessed on: 26 May 2014.

ELECTRICAL ENGINEERING DEPARTMENT OF THE FEDERAL UNIVERSITY OF PARANÁ. **Thermistors - NTC.** 2000. Available at: <http://www.eletrica.ufpr.br/edu/Sensores/2000/brenno/index.html> accessed on: 01 July 2014

EISBERG, Robert; RESICK, Robert. Bohr's model of the atom. In **Physics quantum.** Atoms, molecules, solids, nuclei and particles. 29 reprints. Rio de Janeiro: Elsevier, 1979. Chap. 4, p. 121-167

EISBERG, Robert; RESNICK, Robert. Solids - conductors and semiconductors. In **Quantum physics:** Atoms, molecules, solids, nuclei and particles. 29 reprint. Rio de Janeiro: Elsevier, 1979. Chap. 13, p. 561-608.

SOLAR ENERGY. **Photovoltaic cells.** Available at: <http://energiasolar2012.wordpress.com/celulas-fotovoltaicas/> accessed on: 06 July 2014.

GARAGE ENERGISERS. **Resistors.** Available at: <http://www.engineersgarage.com/tutorials/resistors?page=7> accessed on: 16 June 2014.

EXPLANATORY. **Evolution of the atom:** John Dalton's atomic model. Available at: <http://www.explicatorium.com/CFQ9-Evolucao-atomo.php> accessed on: 01 May 2014.

ENGINEERING FACULTY OF THE PONTIFICAL CATHOLIC UNIVERSITY OF RIO

GRANDE   DO   SUL.   **Transistors.**   2006.   Available   at:
<http://www.feng.pucrs.br/~fdosreis/ftp/elobasicem/Aulas%202006%20II/Aula10/Tran   sistor.pdf>
accessed on: 02 July 2014

GUARATINGUETÁ ENGINEERING FACULTY OF THE UNIVERSITY
SÃO PAULO STATE. **Semiconductors.** Available at:
<http://www.feg.unesp.br/~jmarcelo/restrito/arquivos_downloads/apostilas/eb2/semic
ondut_v1.pdf> accessed on: 16 June 2014.

FERMI NATIONAL ACCELERATOR LABORATORY. **Joseph John Thomson.** Available at:
<http://www.fnal.gov/pub/inquiring/timeline/02.html> accessed on: 06 May 2014.

FERREIRA, José Rincon; BAMPI, Sergio; ALVARES, Lillian. **Semiconductor Information System:**
Strategic importance for attracting investment in the country. In: Ministério do Desenvolvimento,
Indústria e Comércio Exterior; Instituto Euvaldo Lodi/ Núcleo Central. (Org.). O futuro da indústria
de semicondutores: A perspectiva do Brasil. Brasília: MDIC/STI; IEL/NC, 2004, v., p. 193-200.
                    Disponívelem   :<
http://www.redmercosur.org/iepcim/RED_MERCOSUR/biblioteca/ESTUDOS_BRASI
L/BRA_46.pdf> accessed on: 26 April 2014

FILGUEIRAS, Carlos Alberto L.. **Two hundred years of Dalton's atomic theory.** Química Nova na
Escola, São Paulo, n.20, 2004, p. 38-44.

FREIRE, Paulo. **Pedagogy of autonomy:** knowledge necessary for educational practice. 25 ed. São
Paulo: 1996.

GADOTTI, Moacir. Learning with emotion, teaching with joy. In **Boniteza**
of a dream: teaching and learning with meaning. Feevale Rio Grande do Sul: 2003. Chap. 5, p. 45-
56.

GREF - Group for the Redesign of Physics Teaching. Diode and Transistor: Semiconductor materials.
In **Physics 3:** Electromagnetism. 5ed. São Paulo: USP. 2007,
p. 275-297.

CHEMISTRY ATOMS GROUP. **Structure of atoms:** Joseph John Thomson's atomic model   .
                    Available at   :
<http://grupoquimicaatomos.blogspot.com.br/2011_04_01_archive.html> accessed on: 06 June
2014.

HALLIDAY, David; RESNICK, Robert; WALKER, Jearl. **Fundamentals of physics:** optics and
modern physics. v. 4, 8 ed. Rio de Janeiro: LTC, 2011.

INSTITUTE OF PHYSICS OF THE FEDERAL UNIVERSITY OF RIO GRANDE DO SUL. **Bohr**
model   of   the   hydrogen   atom.   Available   at:
<http://www.if.ufrgs.br/~betz/iq_XX_A/modBohr/aModBohrFrame.htm> accessed on: 28 June
2014.

MAINTENANCE   AND   SUPPLIES.   **Bipolar   junction   transistor.**   Available   at:
<http://www.manutencaoesuprimentos.com.br/conteudo/6671-transistor-de-juncao-   bipolar/>
accessed on: 16 June 2014.

MARCONDES, Danilo. The pre-Socratic philosophers. **Initiation to the history of**
philosophy: from the pre-socratics to Wittgenstein. 6ed. Rio de Janeiro: Jorge Zahar, 2001. Chap 2,

p. 30-39.

MECRATONICA ACTUAL. **Thermistors** Available at: <http://www.mecatronicaatual.com.br/educacao/1055-sensores-trmicos-ntc-e-ptc- part-1> accessed on: 17 June 2014.

MELZER, Ehrick Eduardo Martins; AIRES, Aparecida Joanez. **The history of the development of atomic theory:** from Dalton to Bohr. In: VIII Encontro Nacional de Pesquisa em Ensino de Ciências (ENPEC) - I CIEC, 2011, Campinas. Proceedings VIII ENPE - I CIEC, 2011. Available at: < http://www.nutes.ufrj.br/abrapec/viiienpec/resumos/R1348-1.pdf> accessed on: 25 April 2014.

POSITIVE NOTE. **Hydrogen atom and atomic structure.** Available at: <http://www.notapositiva.com/pt/trbestbs/fisica/10_atomo_de_hidrogenio_e_estrutura _atomica_d.htm> accessed on: 26 May 2014

OLIVEIRA, **Ótom Anselmo de; GOMES, Joana D'arc Fernandes. Atomic and molecular architecture.** Natal: UFRN, 2006.

OLIVEIRA, Rebeca Vilas Boas Cardoso; HOSOUME, Yassuko. **Physics Teachers and the Re-elaboration of Practice.** 2008. (Paper/Congress Presentation).

PAULO, Freire. **Pedagogy of autonomy:** knowledge necessary for educational practice. 25 ed. São Paulo: 1996.

PEIXOTO, Eduardo Motta Alves. **SILIUM.** Química nova na escola, n° 14, 2001. Available at: <http://qnesc.sbq.org.br/online/qnesc14/v14a12.pdf> accessed on: 01 July 2014

RADIO AMATEURS. **Transistors.** Available at: <http://www.radioamadores.net/transistores.htm> accessed on: 15 June 2014.

REALE, Giovanni; ANTISERI, Dario. **The "naturalists" or philosophers of "physis".** In **History of Philosophy:** Antiquity and the Middle Ages. São Paulo: PAULUS. 1990. Chap. 2, p. 29-70.

RUSSEL, John Blair. Atoms. In **Química geral.** v.1, 2 ed. São Paulo: Pearson, 1994. ch. 5, p. 205-240.

CHEMISTRY ONLY. **Biographies:** John Dalton. Available at: <http://www.soq.com.br/biografias/john_dalton/> accessed on: 01 May 2014.

CHEMISTRY ONLY. **Electronic layers.** Available at: <http://www.soq.com.br/conteudos/ef/introducaoconstituicao/p1.php> accessed on 26 June 2014.

CHEMISTRY ONLY. **Atomic models:** Niels H. D. Bohr. Available at: <http://www.soq.com.br/conteudos/em/modelosatomicos/p4.php> accessed on: 03 June 2014.

TECMUNDO. **What is a transistor and why is it important for the computer?** 2013. Available at <http://www.tecmundo.com.br/o-que-e/3596-o- what-is-a-transistor-and-why-is-it-important-to-the-computer-.htm> accessed on: 19 June 2014.

TIPLER, Paul Allen; LLEWELLYN, Ralpha. Solid state physics. In **Physics modern.** 5ed. Rio de Janeiro: 2010, p. 262-303.

STATE UNIVERSITY OF WESTERN PARANÁ. **Semiconductor materials.** Available at :
<http://www.foz.unioeste.br/~lamat/downmateriais/materiaiscap15.pdf> accessed on 25 May 2014.

FEDERAL UNIVERSITY OF UBERLÂNDIA. **Materials science and technology:** Semiconductor materials. 2009. Available at
<http://www.ebah.com.br/content/ABAAAeoAYAK/semi-condutores> accessed on: 19 June 2014

UNIVERSITY OF CANTERBURY. **Ernest Rutherford.** Available at:
<http://www.canterbury.ac.nz/rutherford/archive.shtml> accessed on: 06 May 2014

VIANA, H. E. B.; PORTO,P. A. . **The process of developing John Dalton's atomic theory.** Química Nova na Escola. Química Nova na Escola, v. v.7, p. 4-12, 2007.

YOUNG, Hugh D.; FREEDMAN, Roger A. Photons, electrons and atoms. In ______________
Physics IV: optics and modern physics. 12 ed. São Paulo: Pearson, 2003. ch. 38, p. 179216.

yes
I want morebooks!

Buy your books fast and straightforward online - at one of world's fastest growing online book stores! Environmentally sound due to Print-on-Demand technologies.

Buy your books online at
www.morebooks.shop

Kaufen Sie Ihre Bücher schnell und unkompliziert online – auf einer der am schnellsten wachsenden Buchhandelsplattformen weltweit! Dank Print-On-Demand umwelt- und ressourcenschonend produziert.

Bücher schneller online kaufen
www.morebooks.shop

info@omniscriptum.com
www.omniscriptum.com

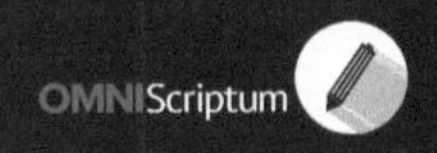

Printed by Books on Demand GmbH, Norderstedt / Germany